20 THINGS YOU DIDN'T KNOW ABOUT

ROCKS AND GEMS

MARIE MORRISON

PowerKiDS press

Published in 2023 by The Rosen Publishing Group, Inc.
2544 Clinton Street, Buffalo, NY 14224

Portions of this work were originally authored by Theresa Morlock and published as *20 Fun Facts About Rocks and Gems*. All new material in this edition was authored by Marie Morrison.

Editor: Greg Roza
Book Design: Tanya Dellaccio

Photo Credits: Cover Victor Moussa/Shutterstock.com; p. 5 Oksana Lyskova/Shutterstock.com; p. 6 DanielFreyr/Shutterstock.com; p. 7 kevin brine/Shutterstock.com; p. 8 Leene/Shutteratock.com; p. 9 Styve Reineck/Shutterstock.com; p. 10 Trygve Finkelsen/Shutterstock.com; p. 11 Laurent CHEVALLIER/Shutterstock.com; p. 12 https://upload.wikimedia.org/wikipedia/commons/b/b3/Soapstone_%28Speckstein%29_-_several_colored_samples.jpg; p. 13 (quartz) PTZ Pictures/Shutterstock.com; p. 13 (diamond) DmitrySt/Shutterstock.com; p. 13 (fluorite) FamStudio/Shutterstock.com; p. 14 Dimj/Shutterstock.com; p. 15 Minakryn Ruslan/Shutterstock.com; p. 16 Rob kemp/Shutterstock.com; p. 17 Bjoern Wylezich/Shutterstock.com; p. 18 DavidNNP/Shutterstock.com; p. 19 Roy Palmer/Shutterstock.com; p. 20 DiamondGalaxy/Shutterstock.com; p. 21 https://upload.wikimedia.org/wikipedia/commons/1/19/Imperial_State_Crown.png; p. 22 REUTERS/Alamy Stock Photo; p. 23 https://upload.wikimedia.org/wikipedia/commons/2/22/Painite2.jpg; p. 24 Evgeny Haritonov/Shutterstock.com; p. 26 The Courage to Travel/Shutterstock.com; p. 27 https://upload.wikimedia.org/wikipedia/commons/f/f1/Oriented_Taza_Meteorite.jpg; p. 29 Kat Om/Shutterstock.com.

Cataloging-in-Publication Data
Names: Morrison, Marie.
Title: 20 Things You Didn't Know About Rocks and Gems / Marie Morrison.
Description: New York : PowerKids Press, 2023. | Series: Did You Know? Earth Science| Includes glossary and index.
Identifiers: ISBN 9781538389768 (pbk.) | ISBN 9781538389782 (library bound) | ISBN 9781538389799 (ebook)
Subjects: LCSH: Geology–Juvenile literature. | Minerals–Juvenile literature. | Precious stones–Juvenile literature. | Rocks–Juvenile literature.
Classification: LCC QE432.2 M67 2023 | DDC 552–dc23

Manufactured in the United States of America

CPSIA Compliance Information: Batch #CWPK23. For Further Information contact Rosen Publishing at 1-800-237-9932.

CONTENTS

JUST A ROCK?

You probably think rocks are nothing special. They're just plain, gray pieces of Earth, right? But rocks can be very cool. They're made of minerals, which are kinds of solid, natural matter that form in the ground. They come in many shapes and colors.

Gems are made of minerals too. They're pieces of rock that people cut and **polish** to use in **jewelry**. They're often crystals, which means they have **structures** arranged in a very regular way with many sides.

Minerals are inorganic, which means that they didn't come from plants or animals.

IT'S IGNEOUS!

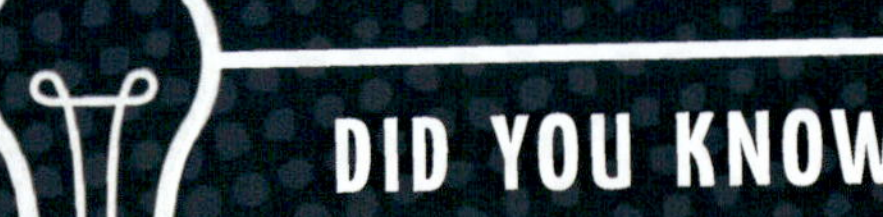

DID YOU KNOW?

Igneous rocks come from lava or magma.

Igneous rocks formed from hot, melted rock. This could be lava, which is melted rock when it comes through the surface of Earth. Or it could be magma, which is melted rock deep under the surface. When this rock becomes solid, it's igneous!

Volcanoes form when melted rock breaks through the crust of Earth.

Granite forms when magma becomes solid when deep underground. It's very hard, and granite rocks can be very old. Basalt, another igneous rock, makes up much of the seafloor.

SEDIMENTARY, MY DEAR

DID YOU KNOW?

Sedimentary rocks can be partly organic.

We usually think of rocks as purely inorganic because minerals are inorganic. However, sedimentary rocks can have bits and pieces of plant or animal remains.

Sedimentary rocks are made of bits of other rocks and matter pressed together underground over time. They often have layers.

Fossils, remains or traces from animals or plants that lived long ago, are most likely to be found in sedimentary rock such as limestone and shale. They're **preserved** as more pieces of rock build up on top of them.

METAMORPHIC MATTERS

DID YOU KNOW?

Metamorphic rocks often start as sedimentary or igneous rocks.

These kinds of rocks become metamorphic when they're **exposed** to high **pressure**, great heat, or some kinds of **fluids** very deep in the earth.

Gneiss, shown here, is a metamorphic rock. Metamorphic rocks can also become different kinds of metamorphic rocks under these conditions.

The word "metamorphic" is related to the word "metamorphosis." This word means a change of physical, or bodily form, such as when a caterpillar turns into a butterfly—or one rock turns into another.

MORE MOHS'

DID YOU KNOW?

The Mohs' scale can tell you how hard or soft a rock or mineral is.

In the early 1800s, a man named Frederich Mohs created a scale that people can use to measure how hard or soft a mineral is. This scale ranges from the mineral talc (1) to the mineral diamond (10).

Soapstone is mainly made of talc, so it's one of the softest rocks. It's a metamorphic rock.

SOAPSTONE

THE MOHS' SCALE

THE MOHS' SCALE INCLUDES INFORMATION ON HOW TO TELL HOW SOFT A MINERAL IS. FOR EXAMPLE, TALC, RATED 1, CAN BE EASILY SCRATCHED BY A FINGERNAIL.

1 TALC CAN BE EASILY SCRATCHED BY A FINGERNAIL

2 GYPSUM CAN BE SCRATCHED BY A FINGERNAIL

3 CALCITE CAN BE BARELY SCRATCHED WITH A PENNY, CAN BE EASILY SCRATCHED WITH A KNIFE

4 FLUORITE CAN BE SCRATCHED WITH A KNIFE, NOT AS EASILY AS CALCITE

5 APATITE CAN BE BARELY SCRATCHED WITH A KNIFE

6 ORTHOCLASE CAN BARELY SCRATCH GLASS

7 QUARTZ CAN SCRATCH GLASS EASILY

8 TOPAZ SCRATCHES GLASS VERY EASILY

9 CORUNDUM CAN CUT GLASS

10 DIAMOND IS OFTEN USED AS A GLASS CUTTER

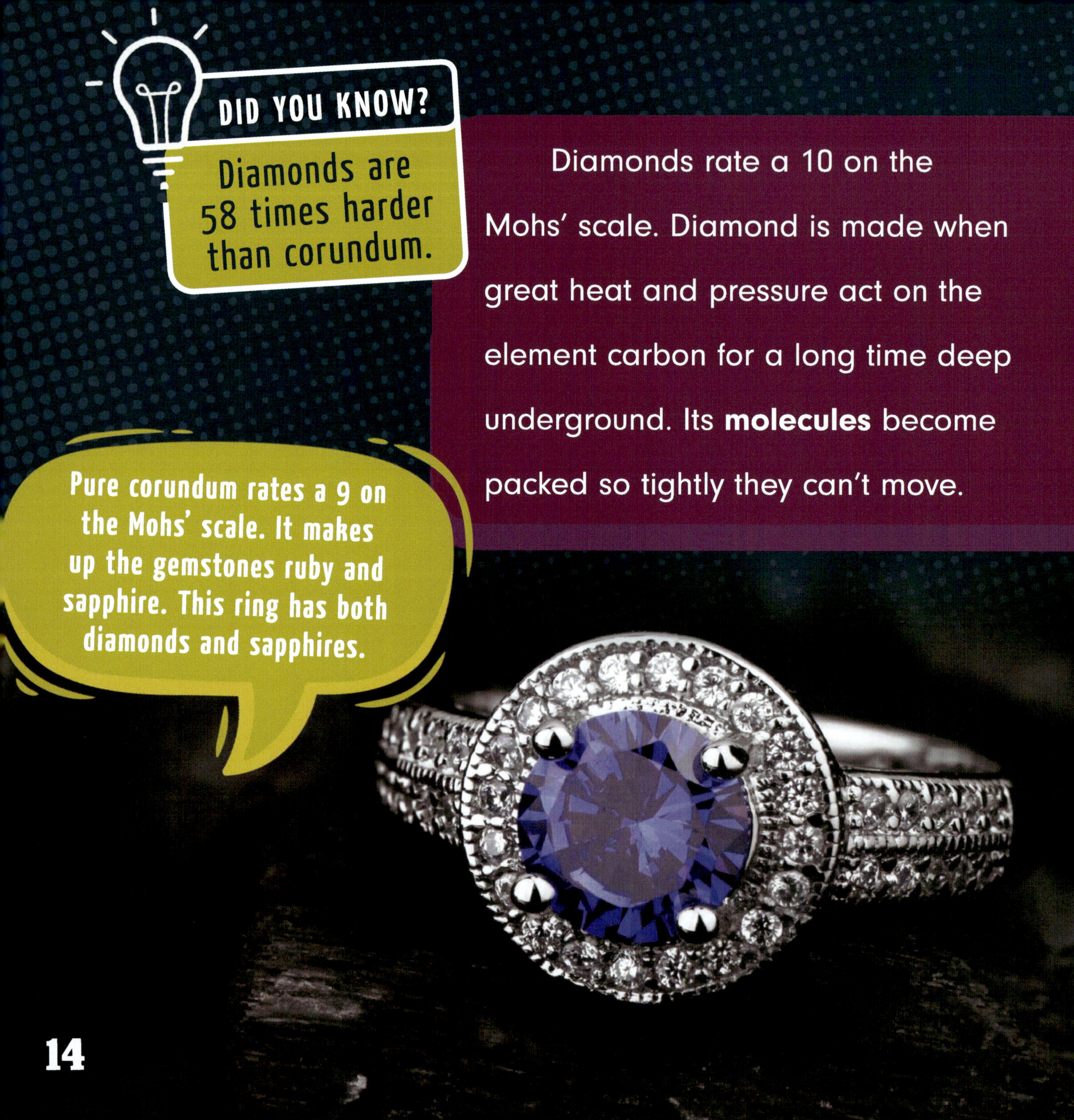

Diamonds rate a 10 on the Mohs' scale. Diamond is made when great heat and pressure act on the element carbon for a long time deep underground. Its **molecules** become packed so tightly they can't move.

DID YOU KNOW?

With the Mohs' scale, direction makes a difference!

Hardness on the scale can depend on what direction you try to scratch a mineral. This is because of the structures that make up crystals. Kyanite, for example, has a 5.5 hardness in one direction and a 7 hardness in another.

FIRE AND WATER

DID YOU KNOW?

Some rocks can float on water.

A kind of rock called pumice is formed when lava and water mix. This cools the lava very quickly and makes its pressure change quickly too. This leaves behind bubbles in the rock. Air in these pockets can make the pumice less **dense** than water.

Pumice rafts are big floating patches of pumice rocks in the ocean. They can be caused by underwater volcanoes.

Rocks are solid and strong, and glass is clear and breakable—right? Not always! Pumice is actually a kind of volcanic glass. Obsidian is another volcanic glass. It's usually shiny and black, and it can have very sharp edges.

SHOWING THEIR COLORS

DID YOU KNOW?

Diamonds come in many colors.

You might think of diamonds as clear, sparkly stones, but they can look very different. There are natural yellow, green, brown, blue, orange, pink, red, black, and violet diamonds, but some of these kinds are very rare—and expensive!

Many colored diamonds get their colors because they have small amounts of other elements. The Hope Diamond, shown, is a blue diamond.

Tourmaline can be multiple shades of many colors. It can be bicolored, which means it has two colors, or multicolored. It can also change color when viewed from different directions.

CROWN JEWELS

DID YOU KNOW?

The weight of a gemstone is measured in carats.

The usual diamond engagement ring in the early 2020s in the United States was about one carat. The Cullinan diamond, found in 1905 in South Africa, was more than 3,000 carats before it was cut!

The Cullinan diamond was cut into nine big gemstones and about 100 smaller ones. The biggest stone is 530 carats!

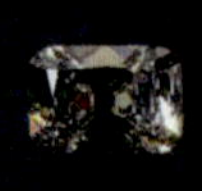
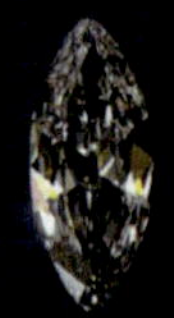
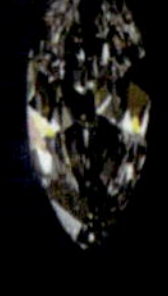

Cullinan II is part of the imperial state crown of the United Kingdom. Queen Elizabeth II wore it after she was crowned queen of the UK in 1953.

DID YOU KNOW?

The biggest Cullinan diamond is the largest clear cut diamond in the world.

This diamond is called the Star of Africa or Cullinan I. The biggest two Cullinan diamonds are part of the crown jewels of the United Kingdom. Cullinan I is set in a scepter, a staff carried by a ruler as a **symbol** of power.

DID YOU KNOW?

Some colored diamonds can cost more than $1 million per carat.

The prices of gems can change based on a number of things, but red and blue diamonds tend to top the price list. They're rare, and good-sized stones with strong color and no flaws are rarer still.

The Oppenheimer blue diamond sold for $57.5 million in 2016. That's well over $1 million per carat.

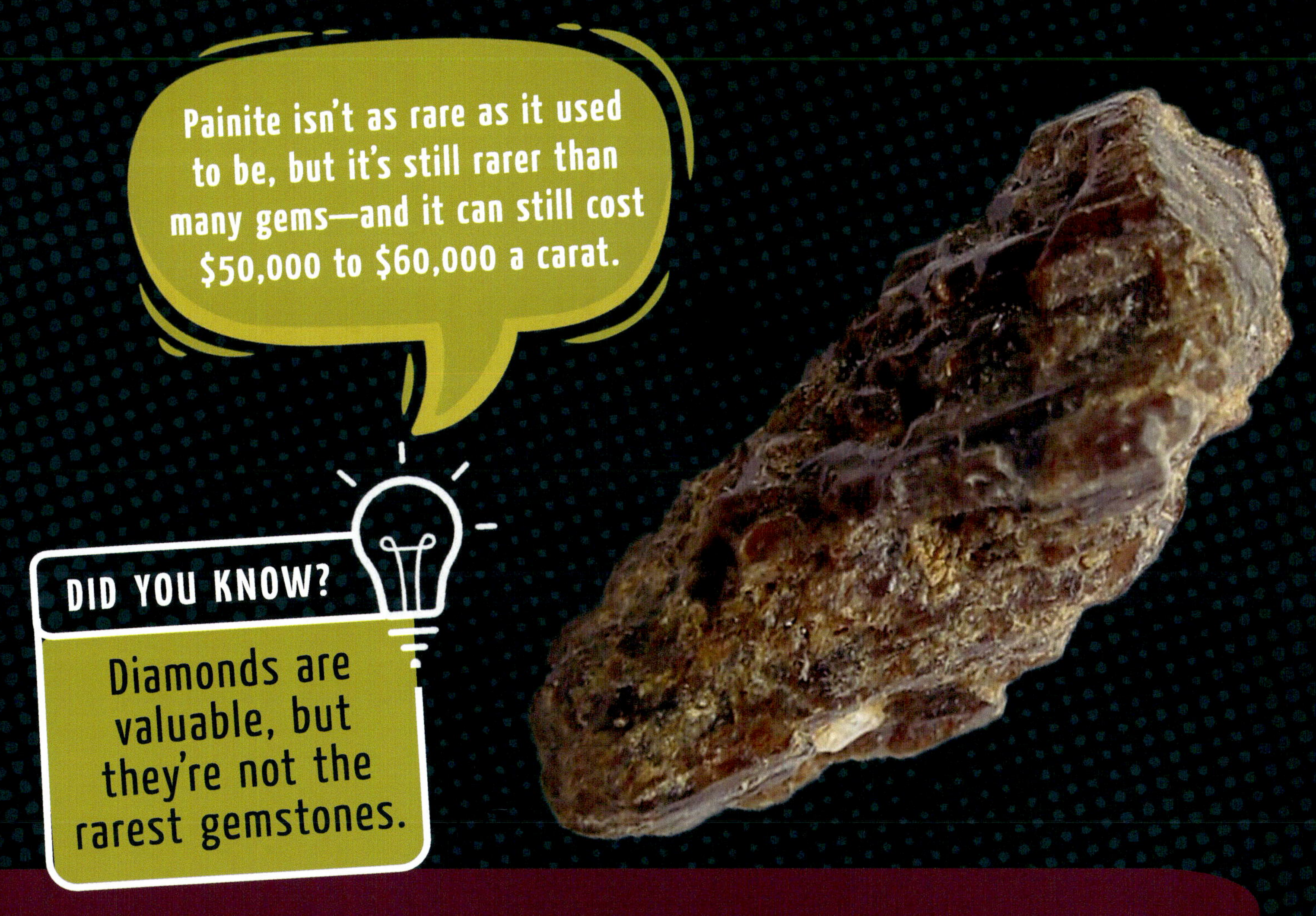

Reports of gemstone rarity vary, but painite is considered one of the rarest. In fact, it's one of the rarest minerals as well. For many years, there was only one example of this reddish-orange or reddish-brown gemstone known.

A RECORD OF HISTORY

DID YOU KNOW?

We can read a lot about Earth's history in the planet's rocks.

Geologists are scientists who study the history of Earth through the geologic, or rock, record. The layers of the earth, or strata, formed as rocks broke apart, formed, and changed. Fossils became part of that rock record too.

Geologists can learn about the climate, or long-term weather patterns, of a past time because of changes in the rock record.

GEOLOGIC TIME SCALE

EARTH IS MORE THAN 4.5 BILLION YEARS OLD. THAT'S A LOT OF TIME TO STUDY! GEOLOGIC TIME IS BROKEN DOWN INTO FOUR MAJOR SPANS OF TIME.

PRECAMBRIAN ERA
4.6 BILLION TO 542 MILLION YEARS AGO

THIS TIME PERIOD STARTS WITH THE FORMATION OF EARTH. IT ENDS ABOUT THE TIME THAT LIFE FIRST CAME INTO BEING ON LAND.

PALEOZOIC ERA
542 MILLION TO 250 MILLION YEARS AGO

THIS ERA STARTED WITH A HUGE EXPLOSION IN THE TYPES OF LIFE ON EARTH. MANY LIFE-FORMS, STARTING WITH PLANTS, MOVED FROM THE OCEANS TO THE LAND. HOWEVER, THE ERA ENDED WITH THE LARGEST MASS EXTINCTION IN HISTORY.

MESOZOIC ERA
250 MILLION TO 65 MILLION YEARS AGO

MANY NEW LIFE-FORMS CAME ABOUT DURING THIS TIME PERIOD. THIS INCLUDES THE DINOSAURS. ANOTHER MASS EXTINCTION ENDED THE ERA.

CENOZOIC ERA
65 MILLION YEARS AGO TO THE PRESENT

MAMMALS TOOK OVER AS THE MAIN TYPE OF LIFE-FORM DURING THIS ERA. HUMANS CAME INTO BEING, CHANGED, AND SPREAD, CHANGING THE PLANET AS THEY WENT.

VISITORS FROM SPACE

DID YOU KNOW?

There are thousands of space rocks on Earth!

Rocks traveling through space are called meteoroids. Once they've entered Earth's **atmosphere**, they're called meteors. If they make it through the atmosphere and hit the ground, they're called meteorites.

More than 50,000 meteorites have been found on Earth.

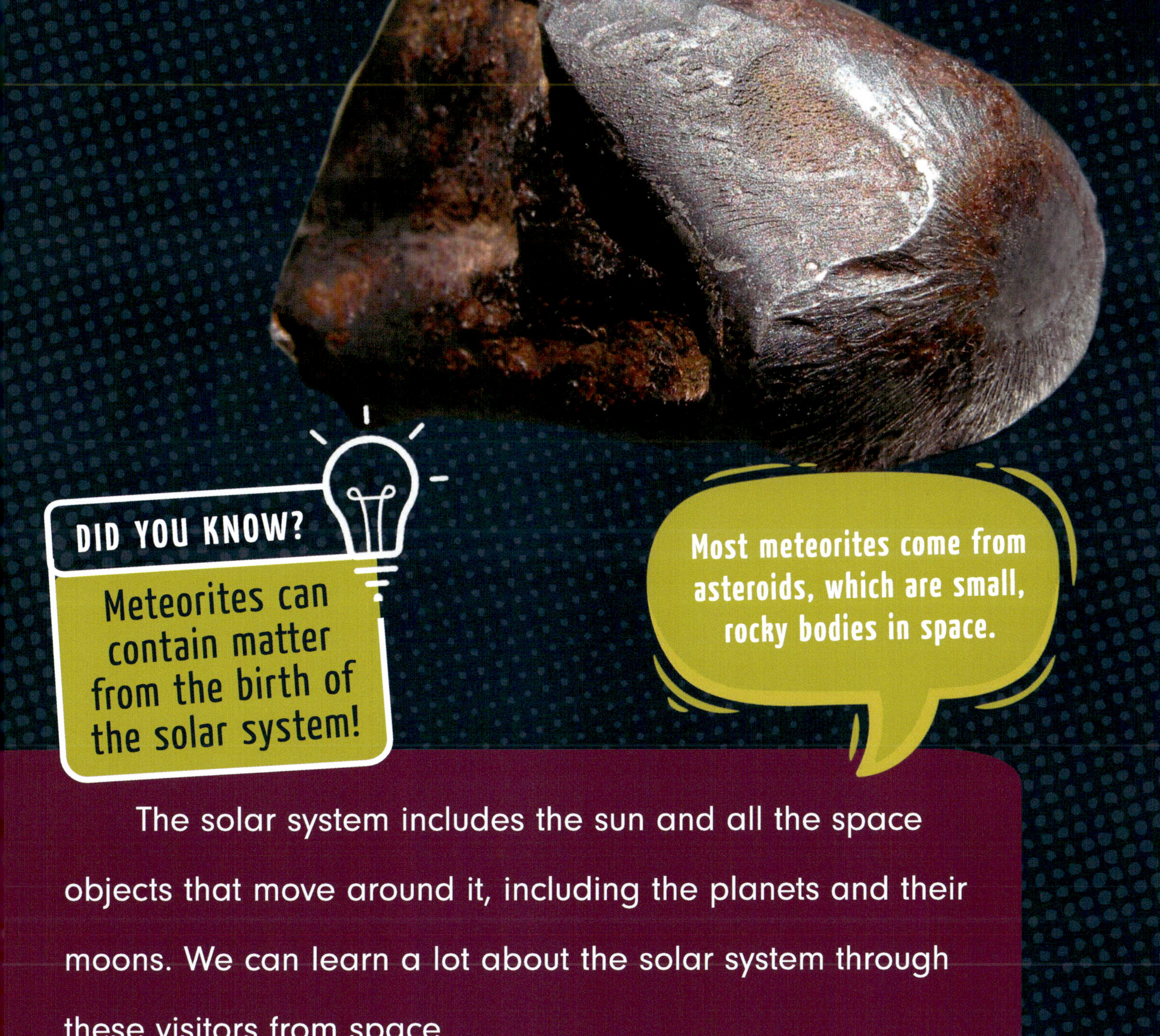

The solar system includes the sun and all the space objects that move around it, including the planets and their moons. We can learn a lot about the solar system through these visitors from space.

BONES OF OUR WORLD

In many ways, rocks are the bones of our world. They can tell us what our planet was like in the past, from the birth of the solar system to climate changes in far more recent history. They can tell us about plants and animals that are long gone and humans that lived before us.

You don't need to be a geologist to learn about rocks. See what kinds of rocks you can find in your own backyard!

What can you learn about the rocks you find around your home?

GLOSSARY

atmosphere: The mass of air that surrounds Earth or another planet.

continent: One of the seven great masses of land on Earth.

dense: Having parts that are close together.

expose: To cause something to be affected by something else.

fluid: A substance that flows freely like water.

jewelry: Objects people wear on their body for decoration, often made of special metals or prized stones.

molecule: The smallest possible amount of something that has all the characteristics of that thing.

polish: To rub something to make it smooth and shiny.

preserve: To keep something in a certain state.

pressure: A force that pushes on something else.

structure: Something that is built by putting parts together in a certain way.

symbol: Something that stands for something else.

FOR MORE INFORMATION

BOOKS

Johnson, Lars W. *Rockhounding for Beginners.* New York, NY: Adams Media, 2021.

Tomaceck, Steve. *Ultimate Rockopedia: The Most Complete Rocks and Minerals Reference Ever.* Washington, D.C.: National Geographic Kids, 2021.

WEBSITES

Gemstones
geology.com/gemstones/
This Geology.com web page has information about and photographs of many kinds of gemstones.

How Are Rocks Made?
wonderopolis.org/wonder/how-are-rocks-made
Wonderopolis offers facts about types of rocks and how they're made.

Meteors and Meteorites
solarsystem.nasa.gov/asteroids-comets-and-meteors/meteors-and-meteorites/
NASA provides a wealth of information about space rocks, including photos and information on meteor showers.

Publisher's note to educators and parents: Our editors have carefully reviewed these websites to ensure that they are suitable for students. Many websites change frequently, however, and we cannot guarantee that a site's future contents will continue to meet our high standards of quality and educational value. Be advised that students should be closely supervised whenever they access the internet.

INDEX